HOME LINKS

Everyday Mathematics®

The University of Chicago School Mathematics Project

Mc
Graw
Hill
Education

Bothell, WA • Chicago, IL • Columbus, OH • New York, NY

The University of Chicago School Mathematics Project (UCSMP)

Max Bell, Director, *Everyday Mathematics* First Edition; James McBride, Director, *Everyday Mathematics* Second Edition; Andy Isaacs, Director, *Everyday Mathematics* Third, CCSS, and Fourth Editions; Amy Dillard, Associate Director, *Everyday Mathematics* Third Edition; Rachel Malpass McCall, Associate Director, *Everyday Mathematics* CCSS and Fourth Editions; Mary Ellen Dairyko, Associate Director, *Everyday Mathematics* Fourth Edition

Authors
Jean Bell, Max Bell, David W. Beer*, Dorothy Freedman, Nancy Guile Goodsell†, Nancy Hanvey, Deborah Arron Leslie, Kate Morrison

* Third Edition only
† First Edition only

Fourth Edition Kindergarten Team Leader
Deborah Arron Leslie

Writers
Kathryn Flores, Gina Garza-Kling, Allison M. Greer, Amanda Louise Ruch, Elizabet Spaepen

Differentiation Team
Ava Belisle-Chatterjee, Leader; Jean Marie Capper

Digital Development Team
Carla Agard-Strickland, Leader; John Benson, Gregory Berns-Leone, Juan Camilo Acevedo

Virtual Learning Community
Meg Schleppenbach Bates, Cheryl G. Moran, Margaret Sharkey

Technical Art
Diana Barrie, Senior Artist; Cherry Inthalangsy

UCSMP Editorial
Molly Potnick, Loren Santow

Field Test Teachers
Rebecca Criollo, Mandy Djikas, Megan DeBetta, Pamela Downing, Tamela Fralin, Laney Frazier, Eric Lester, Priscilla Lindsey, Kari E. Moulton, Nichole Parmley, Mary Rogers

Contributors
Ann E. Audrain, John Benson, Patrick Carroll, Andy Carter, Jeanne Mills DiDomenico, James Flanders, Margaret Krulee, Lila K.S. Goldstein, Barbara Smart, Penny Williams

Center for Elementary Mathematics and Science Education Administration
Martin Gartzman, Executive Director; Meri B. Fohran, Jose J. Fragoso, Jr., Regina Littleton, Laurie K. Thrasher

External Reviewers
The *Everyday Mathematics* authors gratefully acknowledge the work of the many scholars and teachers who reviewed plans for this edition. All decisions regarding the content and pedagogy of *Everyday Mathematics* were made by the authors and do not necessarily reflect the views of those listed below.

Elizabeth Babcock, California Academy of Sciences; Arthur J. Baroody, University of Illinois at Urbana-Champaign and University of Denver; Dawn Berk, University of Delaware; Diane J. Briars, Pittsburgh, Pennsylvania; Kathryn B. Chval, University of Missouri–Columbia; Kathleen Cramer, University of Minnesota; Ethan Danahy, Tufts University; Tom de Boor, Grunwald Associates; Louis V. DiBello, University of Illinois at Chicago; Corey Drake, Michigan State University; David Foster, Silicon Valley Mathematics Initiative; Funda Gönülateş, Michigan State University; M. Kathleen Heid, Pennsylvania State University; Natalie Jakucyn, Glenbrook South High School, Glenview, Illinois; Richard G. Kron, University of Chicago; Richard Lehrer, Vanderbilt University; Susan C. Levine, University of Chicago; Lorraine M. Males, University of Nebraska-Lincoln; Dr. George Mehler, Temple University and Central Bucks School District, Pennsylvania; Kenny Huy Nguyen, North Carolina State University; Mark Oreglia, University of Chicago; Sandra Overcash, Virginia Beach City Public Schools, Virginia; Raedy M. Ping, University of Chicago; Kevin L. Polk, Aveniros LLC; Sarah R. Powell, University of Texas at Austin; Janine T. Remillard, University of Pennsylvania; John P. Smith III, Michigan State University; Mary Kay Stein, University of Pittsburgh; Dale Truding, Arlington Heights District 25, Arlington Heights, Illinois; Judith S. Zawojewski, Illinois Institute of Technology

Note
Too many people have contributed to earlier editions of *Everyday Mathematics* to be listed here. Title and copyright pages for earlier editions can be found at http://everydaymath.uchicago.edu/about/ucsmp-cemse/.

www.everydaymath.com

Send all inquiries to:
McGraw-Hill Education
STEM Learning Solutions Center
8787 Orion Place
Columbus, OH 43240

ISBN: 978-0-02-137954-5
MHID: 0-02-137954-8

Printed in the United States of America.

1 2 3 4 5 6 7 8 9 DOH 19 18 17 16 15 14

Contents

Section 1

Section 1: Family Letter 1

1-3   Counting Steps 3

1-4   Numbers All Around 5

1-7   Family Celebration Math 7

1-9   Count and Seek 9

1-13   Shape and Color Patterns 11

Section 2

Section 2: Family Letter 13

2-3   Triangles at Home 15

2-5   Pocket Problems 17

2-7   Sorting Groceries 19

2-8   Circles at Home 21

2-11   Rectangles at Home 23

2-13   Number Stories at Home 25

Section 3

Section 3: Family Letter 27

3-1   Sorting a Collection 29

3-2   Ten-Coin Toss 31

3-5   Longer or Shorter? 33

3-6   *Simon Says* 35

3-8   Counting and Writing Numbers . . 37

3-9   Line Up 39

3-10   Number-Card Games 41

3-12   *Monster Squeeze* 43

Section 4

Section 4: Family Letter 45

4-1   Attribute Treasure Hunt 47

4-3   Making a Shoe Graph 49

4-5   *Match Up with Ten Frames and Numbers* 51

4-9   Heavier or Lighter? 53

4-10   Measuring Capacity 55

4-11   Counting Fingers 57

4-12   *Top-It with Number Cards* 59

4-13   100th Day Project 61

Section 5

Section 5: Family Letter 63

5-1   Counting on the Number Grid . . 65

5-4   Drawing Favorite Family Shapes . 67

5-5   *I Spy with Shapes* 69

5-8   Teen Partners 71

5-10   Number Stories with Addition . . . 73

5-11   Snack Addition 75

Section 6

Section 6: Family Letter 77

6-1   Reading the Calendar 79

6-2   Comparing Heights 81

6-4   Solid-Shapes Museum 83

6-8   Take-Away Number Stories 85

6-9   Disappearing Snack Trains 87

6-11   *Penny Plate* 89

Section 7

Section 7: Family Letter 91

7-2   "What's My Rule?" 93

7-5   Counting by 10s 95

7-7   Survey Record Sheet 97

7-8   Penny Jar. 99

7-9   Bead Combinations 101

7-11 Class Collection 103

Section 8

Section 8: Family Letter 105

8-2   Modeling Shapes and
      Structures 107

8-4   Counting High and
      Counting On 109

8-6   Grouping Snacks by Tens
      and Ones 111

8-9   Telling Number Stories. 113

8-10 Comparing Ages. 115

8-13 Collections of Number Names . . 117

Section 9

Section 9: Family Letter 119

9-1   Make My Design 121

9-2   Addition and Subtraction 123

9-5   Measuring Objects. 125

9-9   Timing Yourself 127

9-12 Planning for Our Math
      Celebration 129

Dear Families,

Welcome to *Kindergarten Everyday Mathematics,* a curriculum created by the University of Chicago School Mathematics Project. This program is based on research and experience that shows that Kindergartners are capable of far more mathematics learning than was previously believed, provided that the content is presented and explored in age-appropriate ways.

Research also shows that children have more success with written and symbolic mathematics in later grades if they have a Kindergarten curriculum rooted in concrete experience and understanding. Over the course of the year, your child will engage in many hands-on activities related to a range of mathematical topics, including counting, numeration, operations (addition and subtraction), geometry, and measurement. The engaging, playful mathematics activities that children are immersed in throughout *Kindergarten Everyday Mathematics* are designed to help them build a solid foundational understanding of mathematical skills and concepts.

As children participate in *Kindergarten Everyday Mathematics* lessons, they will experience mathematics as useful, enjoyable, and understandable. You can reinforce these experiences at home. Your ongoing involvement with your child around the mathematics that comes up in everyday life will help him or her develop lasting excitement, confidence, and competence in math. You will also periodically receive "Home Links" with activities to do at home that link to those we have done at school.

Ongoing Daily Mathematics Routines

Routines are an important part of daily life in Kindergarten. They provide children with security and predictability, help build classroom community and collaboration, and make aspects of classroom life run more smoothly. They also provide meaningful opportunities to integrate mathematics and other subject areas into everyday activities. We will implement the following classroom routines to provide children with ongoing, real-life opportunities to develop mathematical skills and become mathematical thinkers.

Routine	Children will . . .
Number of the Day	count the days of school and add a new number each day to the Growing Number Line.
Attendance	count the number of children present and absent each day.
Monthly Calendar and Daily Schedule	track the sequence of daily events on a class schedule and track days, weeks, months, and other events on a monthly calendar.
Weather and Temperature	collect, record, graph, and analyze weather observations and temperature measurements.
Survey	collect, record, and graph responses to a "question of the day."

Introduction to Section 1

Kindergarten Everyday Mathematics is organized into 9 sections, each with 13 lessons. Below is information about the main concepts explored in Section 1.

Counting Children will practice the order of number words through counting games, songs, rhymes, and as they do the Daily Routines. They will also count and count out sets of objects. In order to count sets accurately, children must understand the following:

• They say one (and only one) number name for each object, and they cannot skip any objects or count any object twice.

• The last number they count tells the total number of objects in the group.

• The count stays the same regardless of the size, color, shape, or arrangement of the objects or the order in which they were counted.

Developing Number Sense Throughout Section 1, children will be encouraged to notice numbers all around them and to discover the many ways numbers are used. They will also work to represent numbers in many ways.

A poster showing the number 4 in several different ways

Graphing In Section 1, children will also organize data and create class displays to show information about their birthdays and ages.

Display of children's birthdays

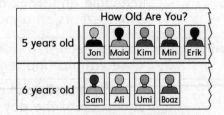

How Old Are You?				
5 years old	Jon	Maia	Kim	Min	Erik
6 years old	Sam	Ali	Umi	Boaz	

Graph showing children's ages

Counting Steps

Count the steps you need to walk from the sidewalk to the front door (or any two places). Try to walk the same distance with fewer steps or with more steps.

Get into the counting habit! When you take a walk, try hopping, skipping, jumping, or sidestepping a certain number of times.

Numbers All Around

Family Note

In this activity, children become more aware of numbers all around them, as well as the varied uses of numbers. Encourage your child to notice numbers in your home. Talk with your child about what the numbers represent and how they are used.

Look for numbers around your home.

Where did you find the most? In your bedroom? In the kitchen? In another room?

Where else did you find numbers?

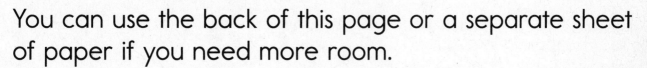

Draw a picture below of some things you found that have numbers.

You can use the back of this page or a separate sheet of paper if you need more room.

Family Celebration Math

Family Note

In school today, we created a class graph to show the months of children's birthdays. Family celebrations provide an excellent opportunity for your child to practice and use his or her developing mathematics skills at home! Try some of these activities with your child before and during family gatherings.

Practice your math when your family gets together for a celebration!

- Count the number of people who are coming. How many are children? How many are adults?

- Set the table with the right number of plates, cups, and napkins. Make sure there is just one of each item for each person.

- Make some place cards, a banner, or other decorations using shapes, patterns, and numbers.

- Help with the cooking. Notice how often you use numbers, counting, and measuring when you cook!

Count and Seek

Family Note

Your child is learning that a quantity of objects is the same regardless of the type of object or the arrangement. For example, although they look very different, all of the following represent 5: 5 blocks stacked up, 5 blocks laying down, 5 big books, 5 little books, 5 fingers on one hand, and 2 fingers on one hand and 3 on the other. Counting out, grouping, and comparing sets of objects of a given number will deepen your child's number sense. In this activity, children practice counting out sets of objects (such as shoes, spoons, stuffed animals, books, and so on) and arranging them in different ways.

Choose a number (1 to 10) as the target number.

One person collects the target number of objects and places them together in a room. The other person tries to find the collection and counts it.

Together rearrange the items in at least two different ways and count them again! You may want to draw a picture of your collection in different arrangements.

Switch roles and play again.

Don't forget to put everything away!

Shape and Color Patterns

Family Note

Your child is learning to identify a *pattern* in a simple arrangement of objects and to predict how the pattern will continue or grow. The concept of predictable patterns is an important part of mathematics.

As your child creates patterns with food or familiar objects, encourage him or her to describe the objects in detail (round, straight, curvy, pointy, and so on) and to name their shapes. Being able to notice and describe these details will help your child learn about shapes and geometry throughout the year.

To make color patterns, dye uncooked pasta by placing the pasta in resealable bags and adding 1 tablespoon of rubbing alcohol and 3 to 4 drops of food coloring for each color you want to create. Shake the bags and then dry the pasta on newspaper.

You can make patterns with food. Use cereal and crackers that have different shapes and colors.

String cereal or pasta (or both) on yarn to make patterned jewelry, or glue them on paper.

Ask a family member to try to figure out your pattern. (Do not eat these patterns!)

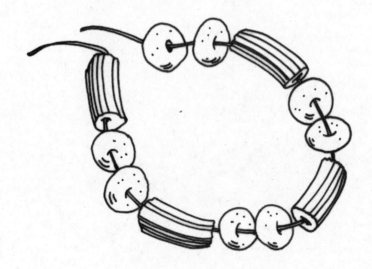

Dear Families,

We are beginning Section 2 in *Kindergarten Everyday Mathematics*. Below is information about the main topics we will learn about during Section 2. We will also continue to explore and practice the concepts and skills we began in Section 1.

Counting and Comparing Sets In Section 2, children will extend early counting experiences to count sets of objects in different arrangements. They will also be introduced to a tool called a *ten frame*. They will use ten frames to see and show numbers in a variety of ways.

The number 4 is shown on a ten frame in two ways.

In *Kindergarten Everyday Mathematics,* children play games frequently to reinforce skills and concepts and develop problem-solving strategies. In Section 2, children will practice counting, matching, and comparing sets of dots by playing *Match Up with Dot Cards* and *Top-It with Dot Cards.*

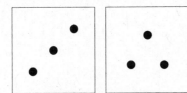

Children find matching sets in *Match Up*.

Children compare sets to see which is greater and which is less in *Top-It*.

Number Stories A "number story" is another name for a word problem or story problem. Early in the year, children solve number stories in a variety of ways, including acting them out and using objects, fingers, and drawings.

"I have 3 red apples and 2 green apples. How many apples do I have in all?"

Shapes Children will make shape collages and explore and describe real-world examples of shapes to help them learn the properties of triangles, circles, and rectangles. They will also learn to recognize the same shape in different sizes and orientations.

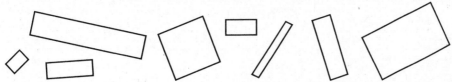

Although the shapes above look different from one another, children learn that all these shapes are rectangles. They also learn that a square is a special type of rectangle!

Triangles at Home

Family Note

We learned about triangles in school today. We learned that triangles can come in lots of different shapes and sizes, as long as they have three straight sides and three vertices (corners). As your child looks for triangles around the house, help him or her find varied examples of triangles in everyday objects and pictures. It is important that children be exposed to a wide variety of examples of each type of shape.

Tell someone at home what a triangle is.

- Look around your house for lots of different triangles. Draw some of them on the back of this page.

- You might also cut out triangles from magazines to make your own triangle collage.

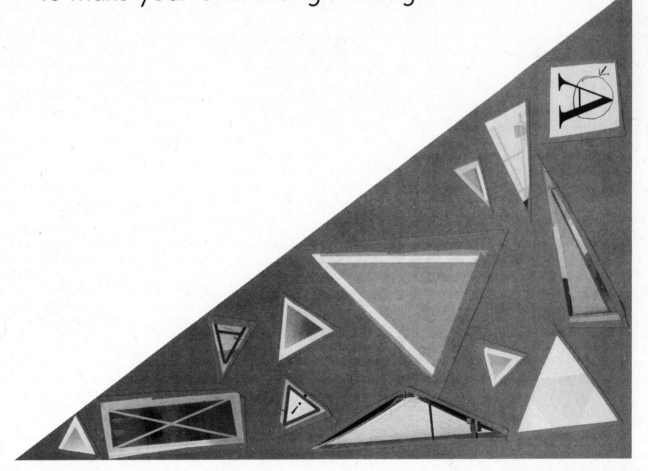

Pocket Problems

Family Note

In school we have begun to explore addition and subtraction using what we call "pocket problems." You can use an envelope, a paper bag, or a cup as your "pocket." Pocket problems allow children to use concrete objects to add and subtract. If your child has difficulty solving the problems mentally, encourage him or her to open the envelope or small paper bag and count the objects.

Use an envelope, a paper bag, or a cup as your "pocket." Do pocket problems with a family member:

- Count aloud as you place a few beans, coins, buttons, or other small objects in your pocket.

- Add two more objects to your pocket.

- Ask your partner to say how many objects there are now without looking inside.

- Pour out the objects and count them to check.

Empty the pocket and repeat with a different number of objects. This time, instead of putting more objects into your pocket, take a few objects out. Ask your partner to say how many objects are still in your pocket.

Sorting Groceries

Family Note

One lesson in each section of *Kindergarten Everyday Mathematics* will be an Open Response and Reengagement Lesson. These lessons provide children with opportunities to solve interesting problems using their own strategies and reasoning. On the first day, children solve an Open Response problem, which is a problem with more than one possible strategy or solution. On the second day, children discuss their work from the first day to reengage with the problem and to learn more about the mathematics involved.

These lessons provide opportunities for children to solve problems that are approachable, but require persistence. Children gain confidence by explaining their thinking and by listening to the explanations of others. They will also see that there is more than one way to solve a problem, which promotes flexible and creative thinking. Ask your child to talk to you about the problems and his or her mathematical thinking throughout the year. You will enjoy seeing your child become a confident problem solver.

In today's lesson, children sorted collections of objects in different ways and explained their groups and sorting rules. Children can apply this type of thinking at home in many ways, including sorting groceries.

Before unpacking a grocery bag, try to guess how many items are inside it. Then count to see how close you were.

- Sort the groceries in the bag into groups. Explain why you put certain items together.

- Can you think of a different way to sort the items?

Circles at Home

Family Note

We learned about circles in school today. As your child looks for circles around the house, help him or her find different examples of circles in everyday objects and pictures. Also help your child notice the difference between true circles and other curved shapes such as ovals or beans.

Tell someone at home what a circle is.

- Look around your house for lots of different circles. Draw some of them on the back of this page.

- You might also cut out circles from magazines to make your own circle collage.

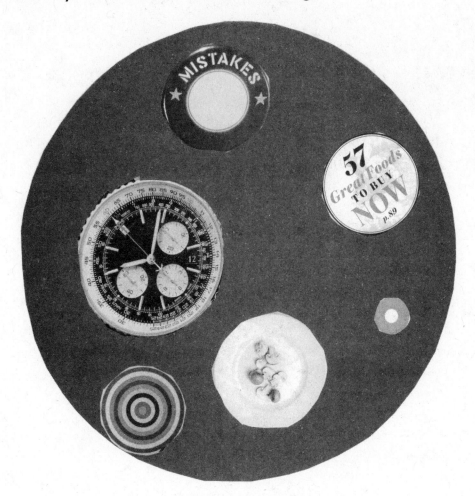

Rectangles at Home

Family Note

We learned about rectangles in school today, including why a square is a special kind of rectangle. (Both have 4 straight sides, with opposite sides the same length. Both also have 4 right angles— vertices that look like "Ls" or "square corners or angles.") As your child looks for rectangles around the house, help him or her find varied examples of rectangles (including squares) in everyday objects and pictures. It is important that children be exposed to a wide variety of examples of each type of shape. Also help your child notice differences between rectangles and other four-sided shapes.

Tell someone at home what a rectangle is. Also explain why a square is a special kind of rectangle.

- Look around your house for lots of different rectangles. Draw some of them on the back of this page.

- You might also cut out rectangles from magazines to make your own rectangle collage.

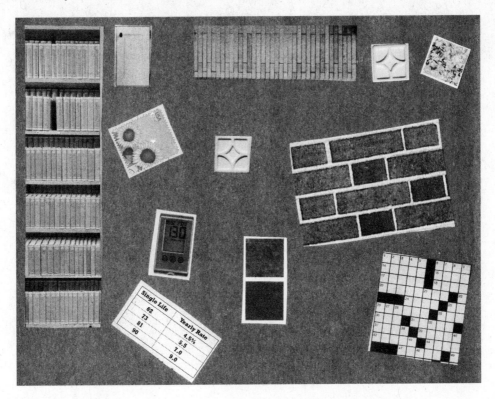

Number Stories at Home

Family Note

We have been telling and solving number stories in school. Number stories help develop children's problem-solving skills and provide a solid foundation for addition and subtraction. At this point in the year, share number stories with your child using informal language and everyday contexts. Focus on numbers within 5 (or within 10 if your child is proficient with smaller numbers). Do not worry about representing the stories with symbols or equations; we will focus on that skill later in the year.

Tell number stories for family members to solve. Ask them how they solved your number stories.

Next have a family member tell a number story for you to solve.

- Show how you can use your fingers, counters, or pictures to model and solve the number story.

- Draw a picture of one of the number stories. Bring it to school to share with the class.

Dear Families,

We are beginning Section 3 in *Kindergarten Everyday Mathematics*. Below is information about the main topics we will learn about during the next few weeks.

Numerals Throughout Section 3, children will make connections between written numbers and what they stand for. For example, the numeral 10 may represent ten fingers, ten stars, or ten spaces on a gameboard. Children will practice writing and interpreting numerals as they create number books and show numbers in many ways.

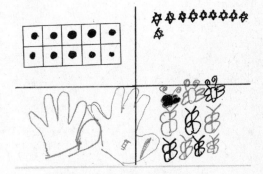

Children show the number 10 in four different ways.

They will also build on their understanding of the number sequence by putting numerals in order, and observing and discussing that each number is exactly one more than the number before it in the counting sequence.

Children will continue playing games to deepen their understanding of numerals. In *Spin a Number* and *Monster Squeeze*, they will practice recognizing and comparing numerals.

| 0 | 1 | | 5 | 6 | | 10 |

In *Monster Squeeze*, children use number relationships (greater/less) to find a mystery number.

Graphing Earlier in Kindergarten, children collected and organized data to create a class birthday display and a graph showing their ages. In Section 3, they will sort pattern blocks to create a graph. (Pattern blocks are used throughout *Kindergarten Everyday Mathematics* to explore shapes and shape combinations.)

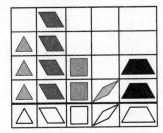

**Children sort pattern blocks by shape and create a graph.
They count and compare totals of each type of shape.**

Sorting a Collection

Family Note

Children love to sort! At school, children sorted blocks by shape and compared the number of blocks in each group. Encourage your child to think of different ways that objects at home can be sorted, such as by color, size, shape, or function. For this activity, help your child find a collection for sorting, such as toy animals or cars, rocks, shells, writing or drawing tools, mixed-shape or mixed-color cereal or pasta, or other items. After sorting, encourage your child to count and to compare the number of items in each group.

Gather a collection of objects.

Sort your objects. (You can sort by, color, size, shape, or another way that is interesting to you.)

When you have finished sorting, count how many objects you have in each group.

- Which group has the **most**?

- Which group has the **fewest**?

- Do any groups have the **same number**?

- **Challenge:** How many more are in one group than another?

Ten-Coin Toss

Family Note

By the end of Kindergarten, children should be able to find number pairs that add to 10, such as 9 and 1, 8 and 2, 7 and 3, and so on. Children are beginning to develop this understanding by playing games with sets of ten and by using ten frames, such as the one below. Do not expect your child to memorize the number pairs at this point. Children will have plenty of practice finding number pairs that add to 10 throughout the year.

Gently toss 10 pennies. Sort the pennies into groups of "heads" and "tails" and put them on the ten frame.

Count the number of heads and the number of tails. You may want to record the numbers you find on the back of this page.

Repeat at least three more times.

Longer or Shorter?

Family Note

Your child is learning about length measurement by comparing objects and describing them as *longer* and *shorter* than other objects. In Kindergarten we focus on direct comparisons of length to prepare children to use measuring tools later. Help your child line up the end of his or her arm (at the longest finger) with the end of the object being compared. This technique will be helpful later when your child learns to line up objects with the end of a ruler or other measuring tool.

Compare the length of your arm (starting from the tip of your longest finger) to objects at home or outside. Which objects are shorter than your arm? Which objects are longer than your arm?

Pretend the picture below is your arm.

Draw one thing you find that is shorter than your arm.

Draw one thing you find that is longer than your arm.

Share your drawings with someone at home.

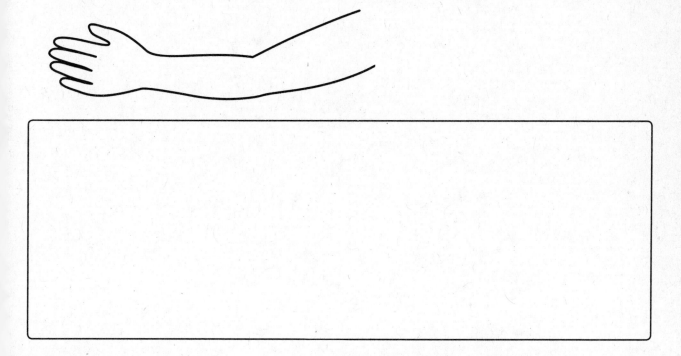

Simon Says

Family Note

In school, children are practicing using positional language to describe objects in the environment. Playing *Simon Says* helps your child learn the meaning of these words by providing opportunities to follow and use positional commands.

Play *Simon Says* with your family. Use positional words such as **above, below, next to, in front of,** and **behind.** Take turns being "Simon" (the leader).

Use clues such as these:

- *Simon says, put your finger **below** your chin.*

- *Simon says, put your foot **next to** your knee.*

- *Simon says, shake your hands **behind** your back.*

- *Wiggle your fingers **above** your head.* (Don't follow this command. Simon didn't say!)

Counting and Writing Numbers

Family Note

At school, your child has been practicing writing the numbers 0–10 to represent sets of objects. Encourage your child to use this new skill by counting items at home and writing numbers to represent his or her findings.

How many of the following things are in your home? Count and write the numbers.

 _____ people

 _____ pets

 _____ beds

 _____ sinks

Draw a picture of your choice and write the number.

Line Up

Family Note

Our class has been learning about the numbers 1–10 and about why their order matters. Connecting numbers to sets and seeing how each number is exactly one more than the number before helps children build important foundations for later arithmetic skills.

Take 1 stuffed animal (or building block or other toy). Cut out the number 1 from a magazine or newspaper or write "1" on a piece of paper. Place the number in front of the toy.

Place a second toy next to the first one. Cut out or write the number that should go in front of the second toy.

Continue until you have a line of 10 toys. Each toy should be labeled with the correct number. Count the toys to check.

Number-Card Games

Family Note

Our class has been learning about numbers and the quantities they represent. Playing games is a fun and effective way to learn about numbers. Many games can be played using a deck of cards. In addition to the games below, your child may also think of other games to play with the cards.

Remove the face cards from a deck of playing cards. Keep the aces and use them to represent the number 1. Use the cards (or your number cards from school) to play these games with a family member or friend.

Nimble Numbers

1, 2, 3, 4, 5!

1. Put your cards in a pile facedown.

2. Pick a card.

3. Choose a movement and do it as many times as the number on the card. For example, if you pick a 5, you may squat or jump 5 times. Count aloud to help you do 1 movement for each number you say.

4. Take turns. Choose a new movement each time!

Line Up

1. Give each player a set with 1 card of each number.

2. Players race to put their cards in order from 1 to 10. You can also play this game by yourself. Ask a family member to time how long it takes you to put your cards in order. Try to beat your best time!

Monster Squeeze

Family Note

Monster Squeeze is a game that reinforces number recognition and the concepts of greater and less. Directions are provided below, but let your child take the lead in teaching you the game.

Materials   Two monsters and a 1–10 number line
Players   2
Object   To guess the mystery number

Directions

1. Player 1 places one monster at each end of the number line, facing the other. The same player chooses a mystery number between 1 and 10 and writes it on a piece of paper.

2. Player 2 guesses a number.

3. Player 1 says whether the number guessed is too low or too high and covers the number with a monster. (The left monster covers the number if the guess was too low. The right monster covers the number if the guess was too high.)

Example: If the mystery number is 6 and the guess is 3, the left monster moves up the number line to cover the 3. If the guess is 8, the right monster moves down the number line to cover the 8.

4. Players keep guessing and moving the monsters until the mystery number is guessed, or "squeezed," between the monsters!

Cut out the monsters and the number line.

Use them to teach someone to play *Monster Squeeze.*

1	2	3	4	5	6	7	8	9	10

Dear Families,

We are beginning Section 4 in *Kindergarten Everyday Mathematics*. Below is information about the main topics we will learn about during the next few weeks.

Counting by 10s and Counting On Children will expand their oral counting skills in Section 4. In addition to counting by 1s through 100, they will count by 10s (10, 20, 30 . . .) and "count on" starting from numbers other than 1.

Exploring Weight and Capacity In previous sections, children practiced describing and comparing *lengths* of objects. In this section, they will explore two other measurable attributes: *capacity* (how much a container can hold) and *weight*.

Children compare the weights of natural objects using a pan balance.

Children fill containers of different shapes and sizes and compare their capacities.

Composing and Decomposing Numbers Children will use sets of objects to explore how numbers can be broken down into combinations of smaller numbers. For example, they may show 5 as a group of 1 blue cube and 4 red cubes, or 2 blue cubes and 3 red cubes, and so on.

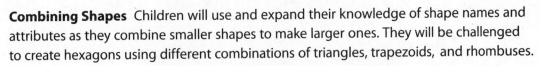

Combining Shapes Children will use and expand their knowledge of shape names and attributes as they combine smaller shapes to make larger ones. They will be challenged to create hexagons using different combinations of triangles, trapezoids, and rhombuses.

Attribute Treasure Hunt

Family Note

Attributes describe the physical characteristics of an object, such as its size, color, and shape. In school, your child has been identifying attributes of blocks and sorting them by common attributes. These skills lay a foundation for later work in geometry and algebra. Plan an attribute treasure hunt at home to help your child practice describing and organizing objects according to multiple characteristics.

Go on an "attribute treasure hunt" with a family member. For example:

- Find a small ball and a large ball.

- Find a thin book and a thick book.

- Find something taller than you and something shorter than you.

- Find as many red objects as you can.

- Find as many round objects as you can.

- Make up your own!

Making a Shoe Graph

Family Note

At school, we made and analyzed a graph of children's favorite colors. At home, children can make a real-object graph by sorting different shoes into categories (shoes with laces, black shoes, fancy shoes, and so on) and lining them up by category. Ask questions to help your child compare the number of shoes in each group. If you do not have enough shoes to sort, your child can use another collection, such as silverware or toys.

Gather the shoes in your house.

Sort the shoes in a way that is interesting to you.

Organize the groups of shoes into lines to make a graph.

Which kind of shoe is the **most** common?

Which kind of shoe is the **least** common?

What other questions can you answer by looking at your graph?

Match Up with Ten Frames and Numbers

Family Note

Ten frames help children see quantities flexibly and break them apart in different ways. They also lay groundwork for learning addition and subtraction facts. Children have played several versions of the game *Match Up* at school. In this version, use the deck of ten-frame and number cards your child brings home to match numerals with ten frames that show that number of dots. Please return the decks so we can continue to play at school.

Materials   Ten-Frame Cards and Number Cards from school
Players   2
Object   To collect the most ten-frame and numeral matches

Directions

1. Place the cards facedown in two rows on a table or on the floor. (Separate the ten-frame row from the number row.)

2. Players take turns choosing one ten-frame card and one number card.

3. If the cards match, the player explains how he or she knows they match and keeps the cards.

Play *Match Up with Ten Frames and Numbers* with someone in your family.

Heavier or Lighter?

Compare the weights of two objects by extending both arms and holding one object in each hand.

Which object is heavier? Which object is lighter?

How did your arms look? Draw a picture to show which object felt heavier and which one felt lighter.

What tools could you use to check which object weighs more?

Try this again with other objects.

Measuring Capacity

Collect some containers that are different shapes and sizes, such as cottage cheese tubs, plastic bottles, and juice containers.

Use the containers to pour water back and forth.

Which container holds the **most**?

Which container holds the **least**?

Do any containers hold about the **same amount**? Which ones? Draw a picture of them.

Counting Fingers

Count all of the fingers in your family.

Count by 10s. (Don't forget yourself!)

Count by 1s to double check.

On the back of this page, draw the fingers of all the people in your family.

Try to write the number of fingers.

(It might be a big number. Ask someone for help if you need it.)

What else could you count by 10s?
(Hint: What do you put in a shoe?)

Top-It with Number Cards

Family Note

Top-It reinforces number recognition and helps children learn to compare two numbers to decide which one is greater or less. (You may remember this game as *War.*)

Materials   Number cards from school or a deck of cards
Players   2
Object   To collect the most cards

Directions

1. Shuffle a deck of cards and then divide it evenly between two players, turning the cards facedown on the table.

2. Players turn over their top cards and read the numbers aloud.

3. The player with the greater number keeps both cards.

Play *Top-It* with someone in your family.

100th Day Project

Family Note

Our class has been adding one number for each school day to our Growing Number Line. The 100th day of school is coming up soon, and it will be a major celebration!

One part of our celebration will be to create a 100 Museum containing collections of 100 things brought in by each child. Children have been thinking about things they might collect, such as a chain of 100 paper clips, a collection of 100 baseball cards, a necklace with 100 beads, or a building made from 100 blocks. We have been talking about ways to count the collections without losing track of the numbers. Making groups of 10 is a good way to count and display the objects.

Children may need a little help gathering materials, but they should be able to do most of the work themselves. Your child can bring in his or her collection as soon as it is ready. We look forward to a rewarding mathematical day!

Create a collection of 100 objects.

Count your objects and arrange them in whatever way you like.

Bring your 100th Day collection to school by

_____ .

Dear Families,

We are beginning Section 5 in *Kindergarten Everyday Mathematics*. Below is information about the main topics we will learn about during the next few weeks.

Teen Numbers So far in this school year, children have worked extensively with numbers 0–10. In Section 5, they will begin to build an understanding of place value by exploring the numbers 11–19. They will use fingers and count to show that these numbers are made up of a group of ten and some additional ones.

The number 16 is represented as "ten and some more ones" with fingers and with counters on a double ten frame.

100th Day of School We have been tracking the number of days we have been in school. On the 100th day, children will count and share their own collections of 100 objects. We will also celebrate with fun math activities such as counting games and 100-themed art, movement, and snack activites.

A collection of 100 stickers

Introduction to Symbols During Section 5, children will be introduced to the equal (=) and addition (+) symbols. They will relate the equal symbol to earlier experiences of showing the same number in many ways. They will use the addition symbol as they act out addition situations with counters. Children will practice using these symbols in games such as *Growing Train,* in which they roll a die labeled + 1, + 2, and + 3 to add to connecting-cube "trains."

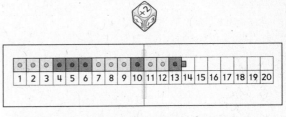

Children become familiar with the addition (+) symbol while playing *Growing Train.*

Shapes Children will continue to explore 2-dimensional shapes by going on a "shape walk" to look for shapes in their environment and by playing *I Spy*. They will draw the shapes they see and practice using positional words to describe them.

"I spy something that is round *next to* the door."

Counting on the Number Grid

Family Note

The 0–100 number grid is a mathematical tool that children can use to help them count, explore number patterns, and develop an understanding of place value. Display the number grid (on the following page) in your home and use it to complete the fun activities below. Return to the number grid frequently to practice the activities with your child.

Look for patterns on the number grid.

- What happens as you move your finger across a row? When you move your finger down a column?

- Where are the smaller numbers? Where are the larger numbers?

Close your eyes and point to a number on the grid.

- Count up *to* that number from 1.

- Or count *from* that number up to 100.

Take turns covering a number with a coin and guessing what the number is. Then say the number right before and right after that number.

Put a coin on the 0 space and roll one or two dice. Move your coin forward the number of spaces shown on the dice, following the order of the numbers on the number grid. Continue until you reach 100!

Invent your own number-grid activities and games!

Counting on the Number Grid (continued)

									0
1	2	3	4	5	6	7	8	9	10
11	12	13	14	15	16	17	18	19	20
21	22	23	24	25	26	27	28	29	30
31	32	33	34	35	36	37	38	39	40
41	42	43	44	45	46	47	48	49	50
51	52	53	54	55	56	57	58	59	60
61	62	63	64	65	66	67	68	69	70
71	72	73	74	75	76	77	78	79	80
81	82	83	84	85	86	87	88	89	90
91	92	93	94	95	96	97	98	99	100
101	102	103	104	105	106	107	108	109	110

Drawing Favorite Family Shapes

Family Note

At school, children are learning to draw shapes. Although drawing is a skill that many Kindergarten children are just developing, it can be used to help them understand the characteristics of shapes. Support your child with this activity by having him or her describe the number of sides and vertices (corners) in the shapes they want to draw. Don't worry if the shapes aren't perfectly drawn.

Ask your family members and friends to tell you their favorite shapes.

Draw and label them below. Describe the shapes to someone.

Draw your favorite shape too!

I Spy with Shapes

Family Note

Your child is learning to recognize and name a variety of shapes. Ask him or her questions about an object's shape whenever the opportunity arises. To broaden your child's knowledge of shapes, be sure to highlight shapes in a variety of sizes and orientations. (Without adult assistance, many children only recognize prototypical shapes, such as a triangle with equal sides or a triangle sitting on one of its bases.) By pointing out and discussing the variety of shapes all around, you will help your child build awareness of geometry concepts and vocabulary.

Play *I Spy with Shapes* with someone. Pick an object that you can see. Give a clue about the shape of the object. Then have the other person guess which object you are describing. Begin with easy clues and then give some harder ones.

Examples:

- "I spy something that is square."

- "I spy something on the wall that is round and has two hands."

- "I spy something that is a rectangle and has rectangular buttons."

Take turns giving clues and guessing.

Teen Partners

Point to and read a number from the strip below.

Ask a family member to show ten fingers.
Use your fingers to complete the number.

Count the fingers all together. Cross the number off the strip and do it again with a new number.

10	11	12	13	14	15	16	17	18	19	20

Number Stories with Addition

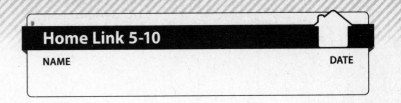

Family Note

Your child has been learning about the addition (or plus) symbol. Share stories that involve putting together or adding to groups of objects to help your child connect the addition symbol to real-world contexts.

Cut out the addition symbol (+).

Take turns telling and solving number stories that use addition. For example: *Two children were on the playground, and three more came to play. How many children were there all together?*

Use pennies or other small objects and the addition symbol (+) to act out, or model, the stories.

Draw or write one of your number stories below.

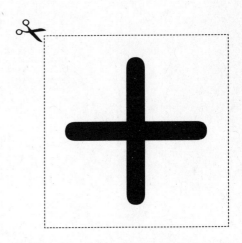

Snack Addition

Family Note

Your child has been learning about adding and about the addition symbol. Cut out the symbol below to make simple addition expressions with snacks. Place the snacks in scattered arrangements to give your child practice counting and organizing scattered sets, which can be difficult.

Put a small number of snacks, such as cereal or raisins, on a table and count them.

Cut out the addition symbol (+) and put it next to the snacks.

Put another group of snacks next to the addition symbol and count them.

Remove the addition symbol and put all the snacks together in one pile. Count the snacks and say how many there are all together. Count again to check.

Repeat with other numbers and snacks.

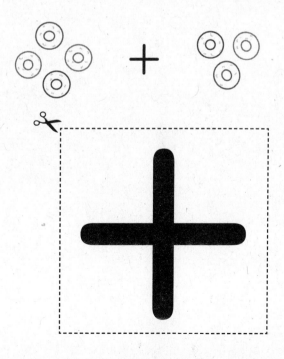

Dear Families,

We are beginning Section 6 in *Kindergarten Everyday Mathematics*. Below is information about the main topics we will learn about during the next few weeks.

Length Measurement Children will use strings to compare their heights to the heights of classroom objects. They will also order objects from shortest to longest. Finally children will explore quantifying length by measuring themselves with stick-on note "units" to see if they are tall enough to go on an imaginary amusement park ride.

2- and 3-Dimensional Shapes Children will learn to describe and name many 3-dimensional shapes, such as cubes, spheres, cylinders, and cones. They will also compare 3-dimensional shapes to one another and to 2-dimensional shapes. In the process, children will notice the many 2-dimensional shapes that form the faces of 3-dimensional shapes and objects.

Addition and Subtraction Situations and Symbols Children learned to use the addition and equal symbols in Section 5. In Section 6, they will use the subtraction symbol to represent "taking away" or "taking apart" situations. Children will make sense of the relationship and differences between addition and subtraction as they solve a variety of number stories and play *Growing and Disappearing Train*.

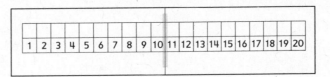

Children use operations symbols and numbers to add and subtract cubes as they play *Growing and Disappearing Train*.

Sorting Children will create and apply rules to sort people or objects into different categories, and they will try to figure out others' sorting rules. For example, while playing *"What's My Rule?" Fishing*, children may determine that they are being sorted by clothing color or the type of shoes they are wearing (or both). Using rules to create and describe categories is an important mathematical skill.

Children discover the "shoe type" rule by noting that all who are asked to stand have shoes that tie.

Reading the Calendar

Family Note

Calendars offer valuable opportunities for children to count and read numbers. We have a calendar routine at school to help us keep track of school events. Have your child help you record and track important family appointments and events on a calendar at home.

Look at your calendar to find answers to the following questions:

- How many days are in this month?

- How many Wednesdays? Fridays? Sundays?

- Are there more Wednesdays or Fridays? Weekdays or weekend days?

- What is today's date?

- How many days are left in this month?

- Are there holidays or special days this month? When are they? Circle or mark them on your calendar.

- Does this month or last month have more days? How many more? How about next month?

Comparing Heights

Family Note

Your child is developing an understanding of measurement by comparing and ordering objects of different sizes from shortest to longest. Build on classroom experiences at home by helping your child record the heights of family members on a doorframe or a large sheet of paper. Measure and mark the same people's heights again in a few months and note whether anyone has grown taller.

Have a family member measure and mark your height. Label it with your initials and the date.

Help measure, mark, and label the heights of other family members.

Compare heights:
Who is tallest?
Who is shortest?
Which family members are closest in height?

Line up your family members in order from shortest to tallest.

Mark and label everyone's heights again in a few months and compare them with today's heights.

Solid-Shapes Museum

Family Note

At school we have been learning about 2-dimensional and 3-dimensional shapes, and children have been noticing shapes all around them. Manipulating, exploring, and discussing 3-dimensional objects helps children learn the names of these objects and builds their spatial sense. Many familiar objects closely resemble 3-dimensional geometric shapes. For example, balls are spheres and dice are cubes. At home, encourage your child to look for and describe 3-dimensional objects to bring to school for our classroom's Solid-Shapes Museum. It is also helpful for children to notice and discuss the 2-dimensional shapes that are part of many 3-dimensional objects, such as the circular faces on the ends of cans and the square faces on dice.

Look around your home for 3-dimensional geometric shapes. Try to find examples like the ones below and describe them to someone.

sphere: ball, globe

cube: dice, square box

cylinder: food can

cone: ice-cream cone, party hat

rectangular prism: cereal box, book

Bring in some objects to add to our classroom's Solid-Shapes Museum.

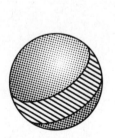

Sphere   Rectangular prism   Cylinder

Take-Away Number Stories

Family Note

Your child has been learning about the subtraction (or minus) symbol. Share stories that involve taking away, or removing, groups of objects to help your child connect the subtraction symbol to real-world contexts.

Cut out the subtraction symbol (−).

Take turns telling and solving number stories that use subtraction. For example: *We made 6 pancakes. Our family ate 4 of them. How many pancakes do we have left?*

Use pennies or other small objects and the subtraction symbol (−) to act out, or model, the stories.

Draw or write one of your number stories below.

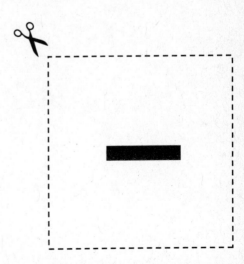

Disappearing Snack Trains

Family Note

Your child has been learning to model "take-away" (subtraction) problems with real objects. For this activity, take all the aces, 2s, and 3s from a deck of cards and shuffle them. Or make 1, 2, and 3 number cards—four cards for each number—out of paper. Your child can help you find or make the cards.

Make two lines of 20 pieces of small snacks, such as fish crackers, cereal, or raisins—a line for you and a line for your partner.

Cut out the subtraction symbols (−) and place them to the right of each snack line.

Players take turns drawing a number card, laying it to the right of their subtraction symbol, and subtracting (eating) that number from their "snack train."

After each turn, compare the trains:
Which has more snacks? Which has fewer?

Play until one player has eaten all their snacks. Shuffle and reuse the cards if you run out.

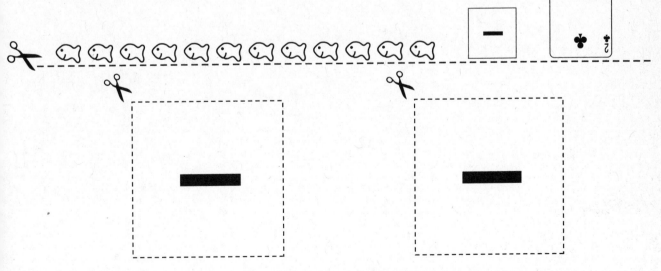

Penny Plate

Family Note

Play *Penny Plate* to help your child practice finding combinations of numbers that add to 10.

Materials   10 pennies, 1 paper plate
Players   2
Object   To figure out how many pennies are hidden

Directions

1. Player 1 turns the plate upside down, hides some of the pennies under the plate, and puts the rest of the pennies on top of the plate.

2. Player 2 counts the pennies on top of the plate and figures out how many pennies are hidden under the plate.

Play *Penny Plate* with a family member. What do you notice about combinations of numbers that add to 10?

Dear Families,

We are beginning Section 7 in *Kindergarten Everyday Mathematics*. Below is information about the main topics we will learn about during the next few weeks.

Addition and Subtraction Strategies Children will solve basic addition and subtraction problems using a counting-on or counting-back strategy. For example, to add 6 and 2: *I can start at 6 and count up 2 numbers: 6 … 7, 8. Six and 2 equals 8.* Children will hop along a large, walk-on number line to model these strategies concretely. They will also add the dots on both sides of dominoes and match the totals with written numerals, noticing different combinations that add to the same number. To develop fluency with addition and subtraction facts within 5, children will learn and play *Dice Addition* in Section 7. (They will learn and play *Dice Subtraction* in Section 8.)

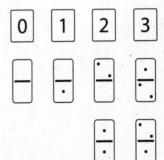

Children add the dots on dominoes and look for different ways to find the same totals.

Collecting and Representing Data During Section 7, children will collect and record data in various contexts. They will ask interesting survey questions of their classmates, and then organize, display, and analyze the response data they collect.

Children will also begin to accumulate data about a class collection. First they will vote on an object the class can collect. Then they will count and record the total as the children bring in objects from home to add to the collection. The collection provides valuable practice in counting to large numbers, counting by 10s and 1s, and organizing and tracking data.

Through these different experiences with data, children will learn about representing and analyzing information in mathematical ways.

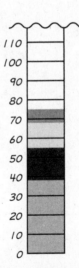

Children represent the number of objects in their class collection using a thermometer-style graph.

Estimation Children will use a reference jar filled with a known number of objects to help them make estimates (or "smart guesses") about the number of objects in a second container. We will revisit this Estimation routine regularly for the remainder of the school year. Children get better at estimation with experience and practice, so look for real-life opportunities for your child to estimate how many people are in a room, snacks are in a bag, flowers are in a garden, and so on! Encourage your child to explain his or her estimate; then count the objects together. Estimation develops number sense and problem-solving skills, so estimate with your child often!

Children use a jar of 10 objects to estimate the number in the second jar.

"What's My Rule?"

Family Note

Your child has been playing a game at school to figure out sorting rules. Use collections of household objects such as coins, toys, stuffed animals, utensils, or a pile of laundry to help your child identify and apply sorting rules.

Materials A collection of similar objects (toys, utensils, clothing, and so on)
Players 2
Object To sort objects and identify a sorting rule

Directions

1. Choose a **rule** (such as "objects that have stripes" or "objects that have wheels").

2. Pick out the objects that follow your rule.

3. Have a family member watch you sort the objects, and then guess your rule.

4. Switch roles and play again!

Play "What's My Rule?" with someone in your family!

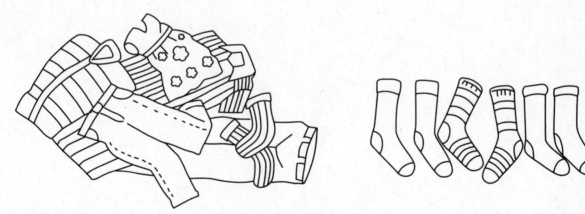

My rule is: Dad's socks!

Counting by 10s

Family Note

In school we are learning to *skip count* by 10s. Counting by 10s can help children count sets of objects more efficiently, and it also helps them recognize and understand number patterns and place value. Look for opportunities to help your child practice counting by 10s.

Show a family member how you can count by 10s to 100. Use the number strip below if you need help.

How many fingers and toes do you have in your home all together? Find the total by counting by 10s!

Look for other things you can count in groups of 10. You can create sets of 10 by placing 10 paper clips, coins, cereal pieces, or other small objects in piles, cups, or plastic bags. Count the collection by 10s.

| 10 | 20 | 30 | 40 | 50 | 60 | 70 | 80 | 90 | 100 |

Survey Record Sheet

Family Note

Observing and collecting data gives children the opportunity to count, to compare numbers, and to think about how numbers can reveal information. Help your child think of a survey question and conduct the survey with family members or friends. (Your child may wish to contact long-distance family members to gather more data.) After your child finishes conducting the survey, have him or her count to find the totals for each category and compare the results.

Conduct a survey among your family and friends. Put possible responses at the top of each column below, and record names in the correct column. Use the back of this sheet or another piece of paper if you need a different kind of chart to record your responses.

Survey Question: _____

Penny Jar

Family Note

A penny jar provides great mathematics opportunities! Have family members add spare pennies at the end of each day. Count the pennies together once a week to reinforce the counting skills we are working on in school. As the penny collection grows, family members can estimate how many pennies are in the jar before counting them. Estimation is not just guessing. It is using what you know to make a "smart guess."

Start a penny jar to collect your family's pennies.

Once a week, estimate how many coins are in the jar:

• Take a small handful of pennies and count them.

• Compare the number in your hand with the number in the penny jar. *How many pennies do you think are in the jar?*

• Count the pennies in the jar and then record the number. *How close was your estimate?*

• *How many pennies do you think will be in your jar next week?* Keep track of how the number changes.

Bead Combinations

Family Note

In school we have been *decomposing,* or grouping, numbers in different ways. For example, the number 6 can be shown as 5 and 1; or 4 and 2; or 3 and 3; or even 3, 2, and 1. Along with this Home Link, your child will bring home a counting loop with beads to practice showing numbers in multiple ways. Please return the loop and beads to school with your child tomorrow.

How many beads are on your counting loop? _____

Take turns with someone at home to find different ways to group the beads. Draw beads and write number sentences to show four of your combinations.

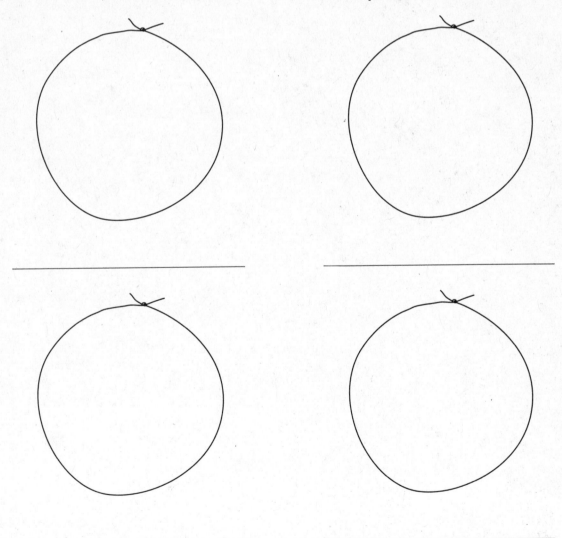

Class Collection

Family Note

Today we began a class collection of objects that we will use at school for number activities such as counting, keeping records, and grouping by 10s. The children voted to collect _____.
Please help your child find items to contribute to our collection. We will build the class collection for at least the next week or two.

Look around your house for _____.

How many do you have? _____

Put them in a bag. Bring the bag to school to add to the class collection.

Bring more of the objects whenever you have them!

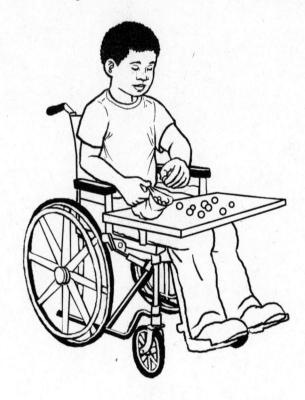

Dear Families,

We are beginning Section 8 in *Kindergarten Everyday Mathematics*. Below is information about the main topics we will learn about during the next few weeks.

Making Ten Children will continue to do activities and play games (such as *Hiding Bears* and *Car Race*) that help them find pairs of numbers that add to 10. In the open response lesson, children will look for as many ways as they can to place 10 birds on 2 wires. As they find combinations that add to 10, children will notice number patterns and prepare for later work with multi-digit addition and subtraction.

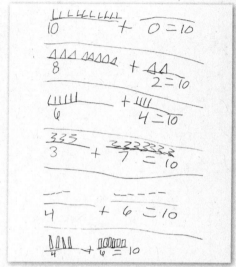

Children find and record many ways to place 10 birds on 2 wires.

Modeling 3-Dimensional Shapes Children will continue their exploration of 2- and 3-dimensional shapes by using toothpicks, marshmallows, and clay to create shapes and then using shape terminology to describe their creations. They will discover that toothpicks are useful for creating shapes with straight edges (such as cubes and prisms), while clay allows for creating shapes with curves (such as cones and cylinders).

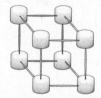

Children build shapes out of marshmallows, toothpicks, and clay.

Adding and Subtracting Children will practice adding and subtracting small numbers by playing games such as *Dice Subtraction* and *Addition Top-It*. As they play, children will recall some sums or differences from memory and will develop and practice quick and accurate strategies for finding others. The goal is for children to develop fluency for sums and differences at least within 5 by the end of Kindergarten. This will also lay the groundwork for fluency with more facts as they move into later grades.

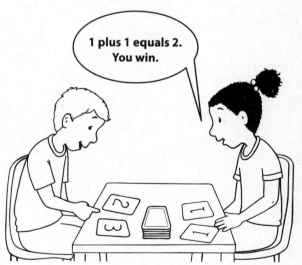

1 plus 1 equals 2. You win.

Measuring Time Children will practice timing classroom activities using steady beats (such as counting "1-Mississippi, 2-Mississippi"). This allows children to practice oral counting to higher numbers in an interesting, meaningful context.

Modeling Shapes and Structures

Family Note

Children can use toothpicks and miniature marshmallows, gumdrops, or small balls of modeling dough as building materials to help develop their understanding of 2- and 3-dimensional shapes. Encourage your child to use the materials to build structures (buildings, vehicles, 3-dimensional designs, and so on) that are made up of common geometric shapes. Help your child learn more about shapes and numbers by talking to him or her about this project. Ask questions such as:

- Are there any squares in your structure?
- How many triangles did you make? How many rectangles?
- Which shapes did you combine to make your creation?
- Do any of your shapes have more toothpicks than marshmallows?
- What 2-dimensional shapes did you make? What 3-dimensional shapes did you make?

Build shapes and structures with toothpicks and marshmallows. (You can use gumdrops or balls of modeling dough instead of marshmallows.)

Build models of **2-dimensional shapes** such as triangles and rectangles. Also build models of **3-dimensional shapes** such as cubes, pyramids, and prisms.

Tell someone at home about your shapes. Then bring one or two of your shapes to school.

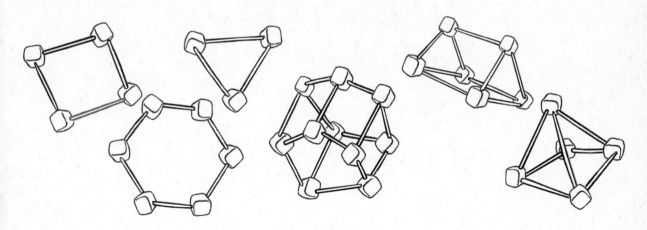

Counting High and Counting On

Family Note

In addition to counting actual objects, children enjoy the rhythm and pattern of reciting numbers in order. As children develop their oral counting skills, they also become aware of the patterns and structure of our number system. Encourage children to count as far as they can, and give subtle hints or prompts to help them count a little higher each time. Children enjoy seeing how high they can go, and they gain a real sense of power when they can start counting from any number.

Practice counting to 100.

Start counting at 1. Then start at another number such as 15, 27, 49, or 62.

Try these counting challenges!

- Count past 100. How far can you go?

- Count down to 0. You may wish to count along with a timer that counts backward.

98, 99, 100... READY OR NOT, HERE I COME.

Grouping Snacks by Tens and Ones

Family Note

Your child is building an understanding of place value by exploring the numbers 10–19. He or she is working to understand that these numbers are composed of ten 1s and some more 1s. (The number 10 has no extra 1s.) This work helps children understand the structure of our base-ten number system and prepares them for more advanced addition and subtraction.

Choose a snack with small pieces, such as cereal or raisins. Count out between 10 and 19 pieces.

Write how many pieces you have. _____

Make a pile of 10 pieces of the snack.
How many pieces are left over? _____

Fill in the double ten frame and write a number sentence to show how many pieces you have.

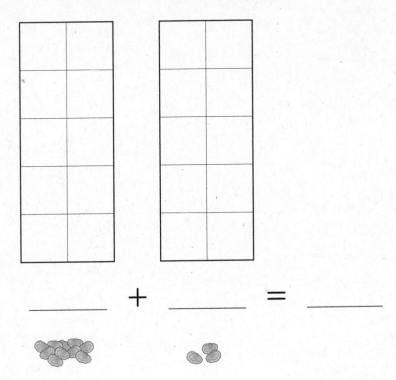

_____ + _____ = _____

Try to count backward to zero as you eat your snack!

Telling Number Stories

Family Note

We have been telling number stories in school and writing number sentences, or equations, to model each story. Help your child become a great problem solver by taking turns telling and solving number stories with him or her. Children especially enjoy number stories that relate to their lives.

Tell number stories about people in your family or places around your home. Start with some "5" stories (stories with 5 as the answer).

For example: *Mom had 6 cans of tuna, but I ate one. How many cans of tuna were left?*

Then have someone in your family tell number stories for you to solve.

Write number sentences for some of your number stories.

Comparing Ages

Family Note

At school we have been comparing and ordering numbers from smallest (least) to largest (greatest). Help your child compare and order the ages of the members of your family. Prompt your child to count up from each number to the next number to check their answers and to practice counting on from numbers other than one.

Draw a picture of your family. Write each person's age.

Write the ages in order from youngest to oldest:

youngest _____ oldest

(least)   (greatest)

Collections of Number Names

Family Note

Today our class created "name-collection posters" by showing numbers in different ways. For example, children showed names for 10 that include 7 + 3; a domino with 5 on one side and 5 on the other; and a tower of 6 red blocks and 4 blue blocks. Creating equivalent names for numbers helps children think flexibly about numbers and recognize that the total stays the same even if the number is broken into groups or represented with pictures, objects, or numbers and symbols. Support your child as he or she shows the number of people in your family in different ways.

How many are in your family? _____
(Include pets and grandparents or other family members if you choose to.)

What are some different ways to group your family members (for example: adults and children, brown eyes and blue eyes, boys and girls, animals and people)?

On the back of this page, write the number of members in your family and show some ways you can group them using pictures, words, and numbers.

EM2007MM_GK_G07_A09_R_0004-657577-A

1 adult + 2 kids + 2 pets   **3 girls + 2 boys**

Dear Families,

We are beginning Section 9 in *Kindergarten Everyday Mathematics*. Below is information about some of our mathematics work during the next few weeks.

Spatial Relationships In Section 9, children will use shape and positional language and develop spatial thinking as they describe a pattern-block design to a partner, who will then try to re-create the design without looking at it. In the Open Response lesson, children will draw maps of the classroom and will later use the maps to locate "hidden treasures." These activities will help children develop spatial reasoning, which is an important aspect of geometry.

Children draw and use maps of the classroom.

Measurement Children will extend and apply the many ways they have learned to describe the sizes of objects as they measure and compare the heights, widths, areas, weights, and capacities of their backpacks. They will also learn to use a pan balance to measure the weight of objects using same-size units, such as paper clips.

Children compare their backpacks along several different size dimensions.

Class Math Celebration Children will end the year by having a two-day math celebration to apply the math skills and understandings they have learned in Kindergarten. On the first day, children will use their number, geometry, and other emerging mathematics skills to write invitations, create decorations, and plan and prepare seating charts, food, and party games. On the second day, children will play their favorite math games, estimate the number of snacks in jars, and celebrate how much math they have learned in Kindergarten!

> Come to our math party!
>
> Friday, June 2
>
> 2:00–3:00 pm
>
> Class 1B
> RSVP to Mrs. Smith

Children apply their math skills to plan and have a math celebration at the end of the year.

Make My Design

Family Note

In school, children used shape names and detailed positional language (such as *next to, above, below, left,* and *right*) to describe and copy each other's shape designs. Use the game below to encourage precise mathematical language.

Materials Stickers with duplicates or different materials with duplicates (such as coins, buttons, beans, pasta, or drinking straws); two pieces of paper; a large book or folder (to use as a divider)

Players 2

Object To re-create pattern-block designs using shape and positional words

Directions

1. Sit next to your partner and place the divider between you.

2. Use your materials to make a design on your paper. Describe the design in detail and have your partner try to make the same design using only your clues.

3. Compare your designs. Do they match?

4. Switch roles and try again!

Play *Make My Design* with someone in your family.

Addition and Subtraction

Family Note

At school, your child has been practicing adding and subtracting small numbers. Children may easily recall some facts but may also need to count on or count back, use their fingers, or count objects to solve other problems. With repeated practice and encouragement, your child will develop more efficient strategies over time. Use the following activities to help your child build *fluency,* the ability to add or subtract quickly and accurately, for sums and differences within 5.

Ask a family member to do these fun addition and subtraction activities with you:

- Call out problems for you to answer with movement. For example: Answer 4 − 1 with 3 claps. Say the equation. Then say the answer and count the motions: 4 − 1 = 3. *1* [clap], *2* [clap], *3* [clap]!

> 4 minus 1 equals 3.
> 1, 2, 3!

- Tell number stories for you to solve. For example: *We had 1 apple and got 3 more. How many apples do we have now?*

1 apple plus 3 more equals 4 apples.

- Take turns rolling two dice and adding or subtracting the number of dots.

2 + 3 = 5

Measuring Objects

Family Note

In school, your child has been learning to describe and compare the height, width, area, capacity, and weight of objects. Support your child by exploring these attributes of objects in your home. Gather containers of various sizes, such as cups, baking pans, baskets, bags, and shoe boxes. Help your child find and use same-size units (such as footsteps, stick-on notes, or paper clips) to measure height and width. Use blocks, beans, snacks, or packing peanuts to fill containers and compare capacities. If you have a scale, your child can weigh the container empty and full.

Choose a container and decide the units you will use to measure.

Measure the height.

It is _____ _____ tall.
      number   unit

Measure the width.

It is _____ _____ wide.
      number   unit

Fill the container.

It holds _____ _____.
          number   unit

It is 8 paper clips wide!

If you have a scale, weigh your container when it is empty and when it is full. Record each weight and compare them. *Why are the weights different?*

You might want to try this again with a new container.

Timing Yourself

Family Note

Your child has been learning to use timers to measure time in seconds. In this activity, your child will explore what he or she can do in 60 seconds (one minute).

Find a timer (a kitchen timer or the timer on a phone) and set it for one minute.

Choose an exercise (for example, sit-ups, leg raises, or jumping jacks). Count how many you can do in one minute. Try a few more kinds of exercises for one minute each. *Which could you do the most times?*

Think of other activities and test whether you can do them in one minute. *Can you write your name five times, empty your backpack, or make a sandwich in 60 seconds?*

Below, draw something you can do in 60 seconds.

Planning for Our Math Celebration

Family Note

We are so excited to use the math skills we learned in school this year to plan for our end-of-year Math Celebration! Please help your child with the task listed below.

Your job for the celebration is:

_____

_____

_____

Planning for Our Math Celebration

Family Note

We are so excited to use the math skills we learned in school this year to plan for our end-of-year Math Celebration! Please help your child with the task listed below.

Your job for the celebration is:

_____

_____

_____